Beer

Beer

A Genuine Collection of Cans

Dan Becker & Lance Wilson

CHRONICLE BOOKS

SAN FRANCISCO

Library of Congress Cataloging-in-Publication Data

Becker, Dan, 1984–
  Beer : a genuine collection of cans / Dan Becker and Lance Wilson.
    p. cm.
  ISBN 978-0-8118-7541-7 (pbk.)
  1. Beer cans—Collectors and collecting—Catalogs. I. Wilson, Lance. II. Title.

  NK8459.B36B42 2011
  663'.42—dc22

                              2010023540

Manufactured in China

Photography and Design by Dan Becker and Lance Wilson

Special thanks to: Katie Barcelona, Michael Bierut, Joe Bottoni, Kristin Cullen, Gina Day, Maureen France, Jayeon Kim, Leah Koransky, Troy Litten, Sandy McGlasson, Chris Miller, Sarah Moffat, Robert Probst, Josh Russo, Therese Russo, Gordon Salchow, Heinz Schenker, Iris Shih, Brian Singer, Allison Weiner, our families and friends, all the companies and designers who created these cans.

10 9 8 7 6 5 4 3

Chronicle Books LLC
680 Second Street
San Francisco, California 94107
www.chroniclebooks.com

Introduction

This book contains a selection of beer cans amassed by beer can collector Josh Russo. Presented alphabetically by brand, the following cans span more than sixty years and thirty countries, many clearly showing their age. They range from simple and crude to delicate and beautiful, and their visual appeal inspired us to photograph them.

One Man's Cans

The year was 1975. Fifth grade was out and summer was in. From baseball to bike riding, you name it, we did it. In our hometown of Cincinnati, Ohio, my friends and I were always up to something, and everything was a competition.

One day, my friend Keith showed me his collection of beer cans. Sure, they looked cool, but he only had fifteen or so, and I knew I could do better. While I had no idea where to start looking, the trash I soon discovered in nearby parks yielded a few of my life-long treasures.

One hot summer day, I ventured out to Ault Park. Not far from the trail I took to get there, I hit the mother lode. There were beer cans everywhere, some caked with dirt, others faded beyond recognition. I came across some trash bags, and loaded them full of the rusty steel and aluminum. It was my first day of searching, and I was hooked.

Over the next few months, my friends and I scoured every park in town. By summer's end, I had amassed over three hundred different beer cans, which I artfully arranged into a giant pyramid in my room. With a little help from my parents, my collection quickly grew larger and more diverse. My dad traveled for business and would pick up cans for me from the cities he visited, while my mom gave me cans she found on the road while riding her bike to the market. During family vacations, I'd fill any extra space in the trunk of our '74 Cadillac with as many beer cans as I could find.

On my fourteenth birthday, I got just what I'd been asking for: a book, appropriately titled *Beer Can Collecting*. It was a stunning visual encyclopedia that included information on where to find beer cans and how to fix damaged ones. From there, my obsession grew, and in the following years, so did my collection.

As I grew older, days spent searching for cans seemed harder to come by. Eventually, my collection found its way into boxes, gathering dust in the basement. My passion for collecting had all but faded when one day my wife asked me what the boxes contained. A bit embarrassed to admit that I was still holding onto what most people considered to be garbage, I quietly replied, "Beer cans." Her suggestion to build shelves to display them on surprised me, and from that point on the thrill of collecting started all over again. Nowadays I dig through online auctions rather than the trash in nearby parks.

I would like to thank my parents, friends, and especially my wife for putting up with me for all these years. Here's to you.

—Josh Russo

Red Top Brewing Co.
Cincinnati, OH | 1950s

Harold C. Johnson Brewing Co.
Lomira, WI | 1950s

F. W. Cook Co.
Evansville, IN | 1950s

Arizona Brewing Co.
Phoenix, AZ | 1960s

Aztec Brewing Co.
San Diego, CA | 1930s/1940s

August Wagner Breweries Inc.
Columbus, OH | 1970s

Acme Brewing Co.
San Francisco, CA | 1940s

Acme

One of the first breweries in California to can their beer was Acme Brewing. This Steingirl can, circa 1957, underwent several design iterations over time that reduced clutter and contemporized the look of the waitress.

Acme Brewing Co.
San Francisco, CA | 1950s

Acme Brewing Co. & General Brewing
San Francisco, CA | 1980s

Cerveza Elaborada por S.A.
El Aguila, Spain | 1970s

Joseph Huber Brewing Co.
Monroe, WI | 1970s

Moosehead Breweries Ltd.
Dartmouth, Canada | 1970s

Peter Hand Brewing Co.
Chicago, IL | 1970s

Pittsburgh Brewing Co.
Pittsburgh, PA | 1970s/1980s

Pittsburgh Brewing Co.
Pittsburgh, PA | 1980s

Archipelago Brewery Co., Ltd.
Petaling Jaya, Malaysia | 1970s

August Schell Brewing Co.
New Ulm, MN | 1970s

August Schell Brewing Co.
New Ulm, MN | 1970s

ANOKA
HALLOWE
FESTIVAL
BEER

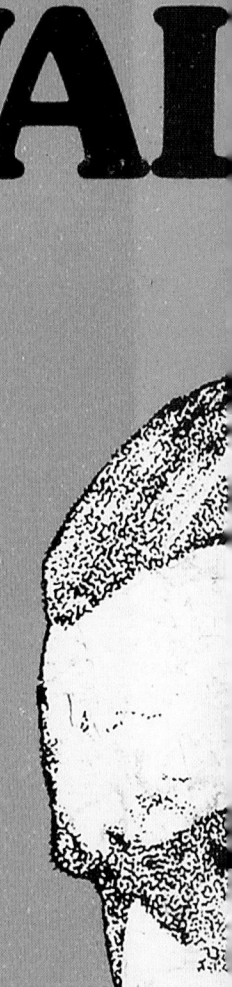

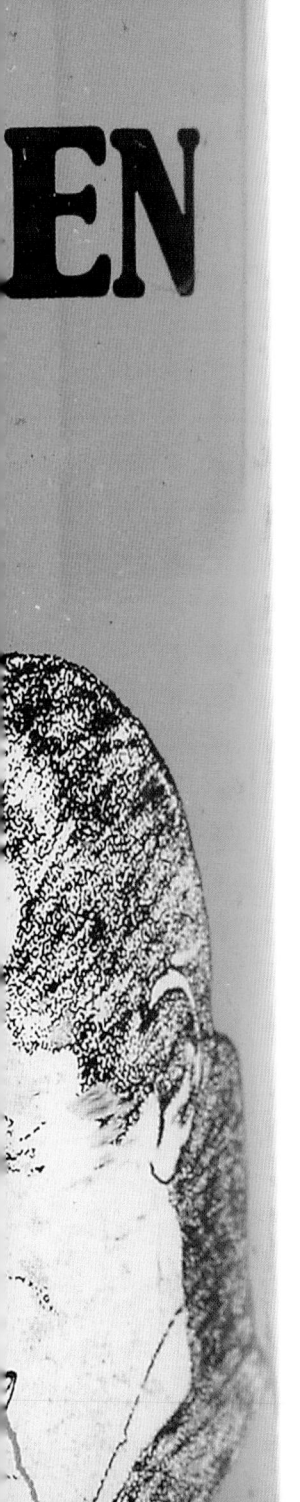

Anoka Halloween Festival Beer

Referred to as "The Halloween Capital of the World," Anoka, Minnesota, is the hometown of this holiday brew. Produced for the 1979 Halloween festival by the August Schell Brewing Company, this blood-red-colored beer featured a vampire on its label.

August Schell Brewing Co.
New Ulm, MN | 1970s

Asahi Breweries
Tokyo, Japan | 1970s/1980s

Atlantic Co.
Atlanta, GA | 1940s/1950s

Pittsburgh Brewing Co.
Pittsburgh, PA | 1970s/1980s

the
Avenue
fine pilsner
BEER
12 FL. OZ.

August Schell Brewing Co.
New Ulm, MN | 1980s

PACKED AT
CONTENTS 12 FLUID OZ. — WITH
PURITY
BODY
FLAVOR
© 1941 TRADE MARK REG U.S. PAT. OFF.
EXPORT
BREWED AND PACKED BY
P. BALLANTINE & SONS
NEWARK, N.J.
LIGHT

Ballantine's Export

During World War II, the scarcity of metal forced many U.S. breweries to stop producing cans. This required many beers to be bottled rather than canned, with the exception of those packaged for soldiers. This Ballantine's Export was colored to avoid reflecting light at night, helping to prevent a soldier's location from being revealed.

P. Ballantine & Sons
Newark, NJ | 1940s

Red Top Brewing Co.
Cincinnati, OH | 1940s/1950s

Walter Brewing Co.
Eau Claire, WI | 1970s

The Lion Brewery, Inc.
Wilkes-Barre, PA | 1980s

Joseph Huber Brewing Co.
Monroe, WI | 1970s

Becker Products Co.
Ogden, UT | 1960s

Berghoff Brewing Co.
Fort Wayne, IN | 1950s

Pabst Brewing Co.
Milwaukee, WI | 1960s

Billy

Capitalizing on the attention he received as the brother of then–United States President Jimmy Carter, the eccentric Billy Carter endorsed this beer of the 1970s.

The West End Brewing Co.
Utica, NY | 1980s

Garden State Brewing Co.
Hammonton, NJ | 1970s

The Leisy Brewing Co.
Cleveland, OH | 1950s

Carling O'Keefe Breweries
Toronto, Canada | 1970s

Black Horse Brewery of New Jersey
Trenton, NJ | 1970s

Brewing Corp. of America
Cleveland, OH | 1950s

Carling Brewing Co.
Baltimore, MD | 1970s

Blatz Brewing Co.
Milwaukee, WI | 1930s

Blatz Brewing Co.
Milwaukee, WI | 1940s

North Bay Brewing Co.
Santa Rosa, CA | 1940s/1950s

August Schell Brewing Co.
New Ulm, MN | 1980s

Bohemian Breweries Inc.
Spokane, WA | 1960s

Joseph Huber Brewing Co.
Monroe, WI | 1970s

The Burger Brewing Co.
Cincinnati, OH | 1960s

Bosch Brewing Co.
Houghton, MI | 1950s

Boston Light Ale

Introduced on a radio show in the early 1940s, this Boston Light Ale claimed to be from the oldest brewery in America. The label featured an illustration of the Boston lighthouse—a national historic landmark and the first lighthouse in North America.

Boston Beer Co.
Boston, MA | 1930s/1940s

Companhia Cervejaria Brahma
Rio de Janeiro, Brazil | 1970s

Independent Milwaukee Brewery
Milwaukee, WI | 1940s/1950s

Moninger Brauerei
Karlsruhe, Germany | 1970s

Rice Lake Brewing Co.
Rice Lake, WI | 1970s

Horlacher Brewing Co.
Allentown, PA | 1970s

Brockert Brewing Co., Inc.
Worchester, MA | 1930s/1940s

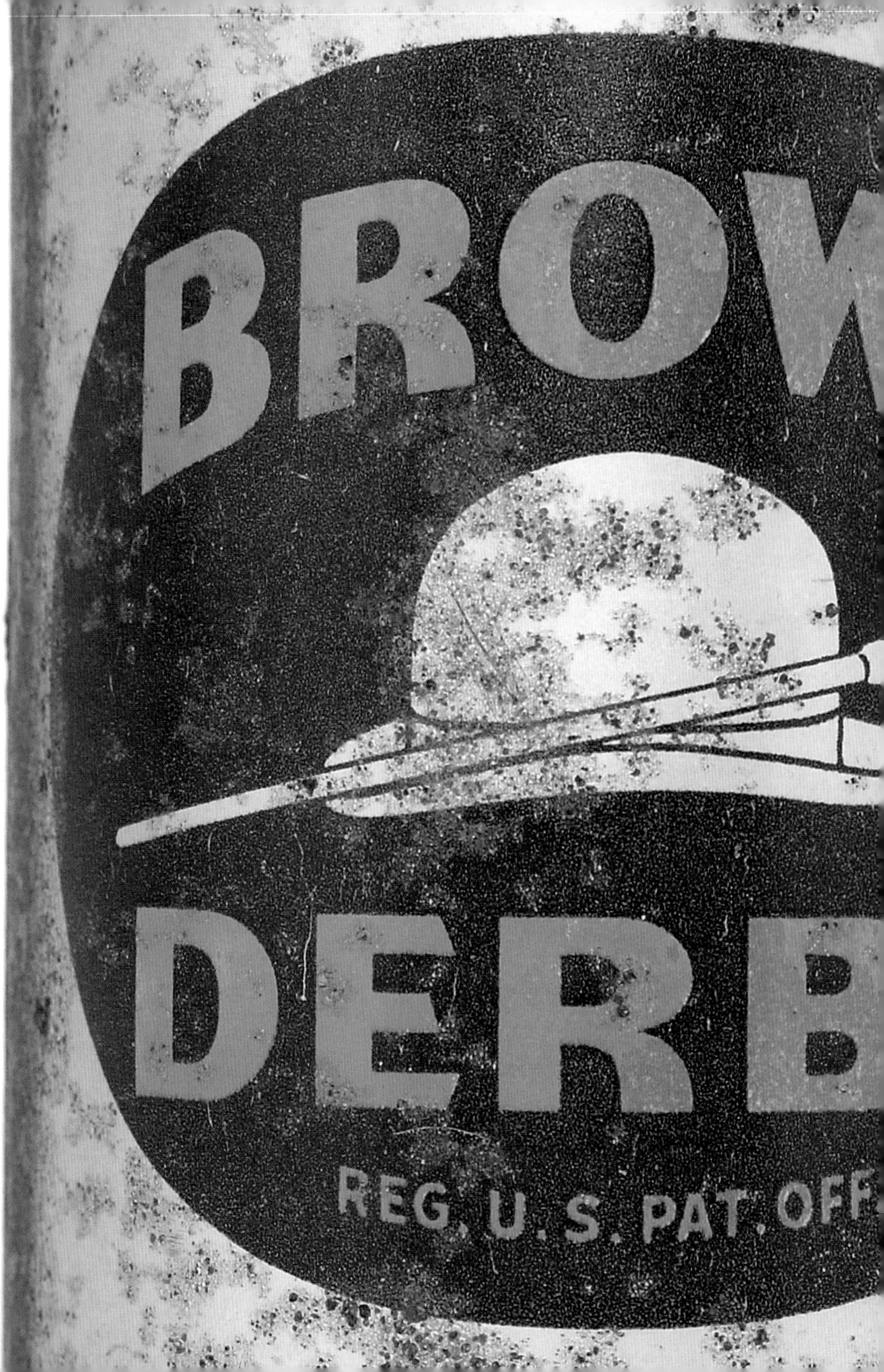
BROW
DERB
REG. U.S. PAT. OFF

Brown Derby

Produced in the 1930s as the house brand for supermarket chain Safeway, Brown Derby beer shared the same name as a then-famous restaurant in Los Angeles, California. The restaurant's owner filed a lawsuit for copyright infringement, forcing the cans to be redesigned.

Humboldt Malt & Brewing Co.
Eureka, CA | 1930s

Century Brewery Co.
Norfolk, VA | 1950s

General Brewing Co.
Vancouver, WA | 1980s

GOOD OLD

Bruck's

SINCE 1856

BEER

Brewed and Packaged by
THE BRUCKMANN CO.· CINCINNATI, OHIO

The Bruckmann Co.
Cincinnati, OH | 1940s

The Bruckmann Co.
Cincinnati, OH | 1940s

12 FL. OZ.

SINCE 1838

Buckey

Pilsener

BEER

Buckeye Pilsener

Seen dressed in a waiter's outfit carrying a tray full of beer, "Bucky" was the mascot for the Toledo, Ohio-based Buckeye Brewing Company. Named after the Buckeye trees common throughout the state, the brewery was in operation for nearly 130 years.

The Buckeye Brewing Co.
Toledo, OH | 1940s/1950s

Buckhorn Brewing Co.
St. Paul, MN | 1960s

Buckhorn Brewing Co.
St. Paul, MN | 1970s/1980s

TRADE MARKS   REG. U. S. PAT. OFF.

Budwei

Budweiser

One of the oldest known Budweiser cans, this design dates back to the 1930s. The bald eagle was first used by the company in 1872, and remains an icon of the brand to this day.

Anheuser-Busch Inc.
St. Louis, MO | 1930s

Anheuser-Busch Inc.
Newark, NJ | 1960s

Anheuser-Busch Inc.
St. Louis, MO | 1960s

Buffalo

Like many others, Buffalo Brewing founder Henry Grau named his brewery after a U.S. city. In this case, however, the brewery resided in Sacramento, California, rather than Buffalo, New York—over 2,000 miles away.

Buffalo Brewing Co.
Sacramento, CA | 1950s

Bull Dog Lager BEER BY ACME
Brewed to a man's taste

Acme Brewing Co.
San Francisco, CA | 1950s

The Burger Brewing Co.
Cincinnati, OH | 1940s

The Burger Brewing Co.
Cincinnati, OH | 1940s

The Burger Brewing Co.
Cincinnati, OH | 1940s/1950s

Burgermeister Brewing Co.
San Francisco, CA | 1940s

Burgermeister Brewing Co.
San Francisco, CA | 1960s

Burgie Brewing Co.
San Francisco, CA | 1970s/1980s

The Burkhardt Brewing Co.
Akron, OH | 1940s/1950s

California Brewing Co.
San Francisco, CA | 1950s

Canadian Ace Brewing Co.
Chicago, IL | 1940s/1950s

Standard Brewing Company of Scranton
Scranton, PA | 1940s/1950s

Carlsberg Breweries
Copenhagen, Denmark | 1970s

Tasmanian Breweries Pty. Ltd.
Hobart, Australia | 1970s

Casey's Lager

Former Philadelphia Phillies centerfielder Richie Ashburn adorns this Casey's Lager beer can, one in a series of cans that featured major league baseball stars.

Valley Forge Brewing Co.
Philadelphia, PA | 1970s

South African Breweries
Johannesburg, South Africa | 1970s

Terre Haute Brewing Co., Inc.
Terre Haute, IN | 1940s/1950s

Joseph S. Pickett & Sons, Inc.
Dubuque, IA | 1970s

Champale Inc.
Trenton, NJ | 1970s

Oshkosh Brewing Co.
Oshkosh, WI | 1940s/1950s

Jacob Leinenkugel Brewing Co.
Chippewa Falls, WI | 1970s

Jacob Leinenkugel Brewing Co.
Chippewa Falls, WI | 1970s

Christian Moerlein
Cincinnati, OH | 1970s/1980s

IMPORTED
FROM CANADA

Cinci
LAGER BEER

CONTENTS 11½ U S FLUID OZS

Carling O'Keefe Breweries
Toronto, Canada | 1970s

Jacob Schmidt Brewing Co.
St. Paul, MN | 1950s

Eastern Brewing Co.
Hammonton, NJ | 1960s

Cold Spring Brewing Co.
Cold Spring, MN | 1940s/1950s

Carling National Breweries
Baltimore, MD | 1970s/1980s

Carling Brewing Co.
Baltimore, MD | 1970s/1980s

SPRING BREWIN

old Spring

LAGER BEER

NTENTS 12 FLUID OUNC

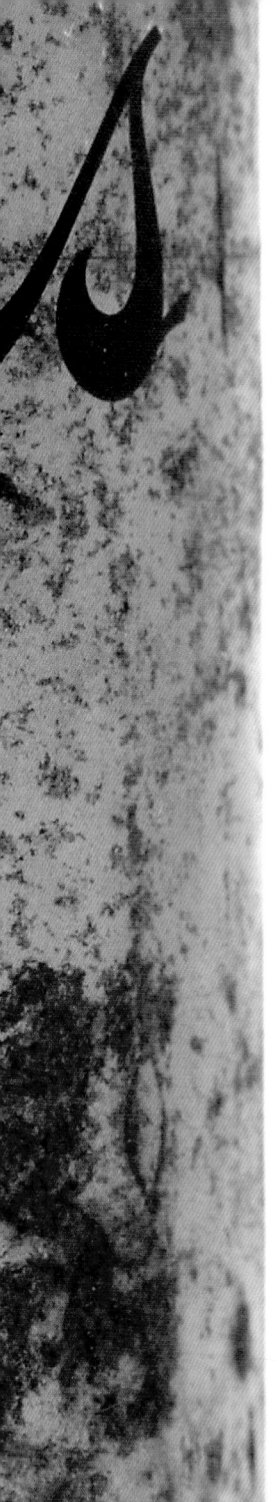

Coors Banquet

The design of this Coors banquet can, featuring an illustration of flowing spring water, remained virtually unaltered from the 1950s to the 1970s.

The famous "Rocky Mountain Spring Water," discovered by Adolph Coors, was found in Clear Creek Valley near Golden, Colorado, where the company was founded and remains to this day.

Adolph Coors Co.
Golden, CO | 1950s

Adolph Coors Co.
Golden, CO | 1970s/1980s

CONTENTS 12 FLUID OUNCES

Brewed with Spring Water

CC

Copper Club

PILSNER

BEER

BREWED AND FILLED BY
A. HAAS BREWING COMPANY, HANCOCK, MICHIGAN

A. Haas Brewing Co.
Hancock, MI | 1940s/1950s

Cerveceria Corona Inc.
San Juan, Puerto Rico | 1970s

Courage Ltd.
Lewes, England | 1970s

Dawson's Brewery Inc.
New Bedford, MA | 1930s

Dawson's Brewery Inc.
New Bedford, MA | 1930s/1940s

The Christian Diehl Brewing Co.
Defiance, OH | 1940s

Dixie Brewing Co.
New Orleans, LA | 1970s

Schoenhofen-Edelweiss Co.
Chicago, IL | 1950s

Allied Breweries
Burton, England | 1970s

Dominion Breweries Ltd.
Auckland, New Zealand | 1970s

Drewrys Ltd.
Chicago, IL | 1960s

G. Heileman Brewing Co.
Newport, KY | 1980s

Du Bois Brewing Co.
Du Bois, PA | 1940s/1950s

Du Bois Brewing Co.
Pittsburgh, PA | 1960s

Duquesne Brewing Co.
Philadelphia, PA | 1970s/1980s

Duquesne Brewing Co.
Pittsburgh, PA | 1940s/1950s

Duquesne Brewing Co.
Pittsburgh, PA | 1940s/1950s

Duquesne Brewing Co.
Pittsburgh, PA | 1940s/1950s

Dutch Treat Brewing Co.
Phoenix, AZ | 1970s/1980s

© COPYRIGHT 1947
L. A. BREWING CO.
EASTSID
TRADE MARK REG.
U. S. PAT.
BEER

Eastside

Founded in 1907, the Los Angeles Brewing Company's location on the east side of the Los Angeles River inspired the name of this beer. With the onset of Prohibition, Los Angeles Brewing converted Eastside into a near beer—a malt beverage containing little or no alcohol.

Because the process of making near beer was similar to producing the alcoholic variety, the brewery was able to easily convert back to shipping alcoholic beer the minute Prohibition expired on April 7, 1933.

Los Angeles Brewing Co.
Los Angeles, CA | 1940s

Ebling Brewing Co., Inc.
New York, NY | 1940s/1950s

El Rey Brewing Co.
San Francisco, CA | 1930s/1940s

Essling

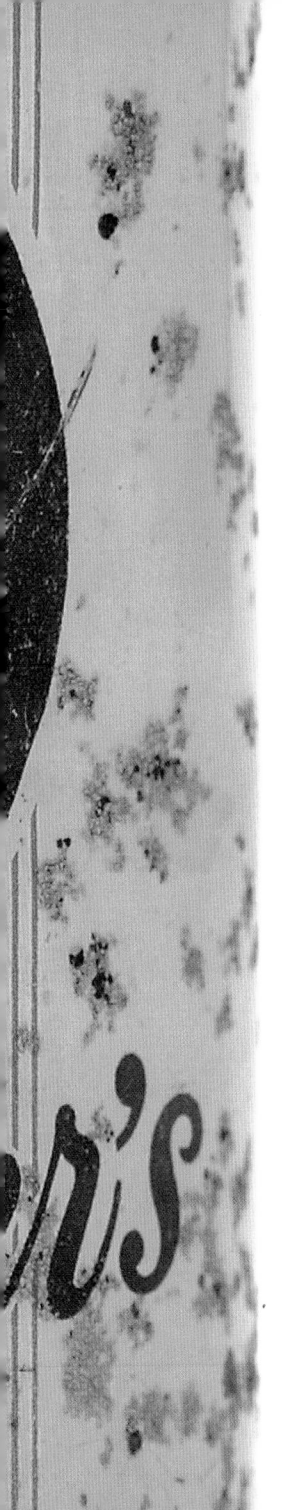

Esslinger's Little Man Ale

Operating out of Philadelphia, Pennsylvania, Esslinger Brewing used the quirky waiter seen on this can as its logo throughout the 1940s and '50s. The mascot, known as "Little Man," was a charming and popular icon used in numerous marketing materials.

Esslinger's Inc. Brewery
Philadelphia, PA | 1940s/1950s

EXTRA PALE

Eureka Beer

Eagle Brewing Co.
San Francisco, CA | 1930s/1940s

Joseph Huber Brewing Co.
Monroe, WI | 1970s

Falstaff Brewing Co.
St. Louis, MO | 1940s/1950s

CONTENTS 12 FL. OZ.

FALSTAFF

FALSTAFF

BEER

The Choicest Product of the Brewer's Art

BREWED AND
FILLED BY FALSTAFF BREWING CORP., ST. LOUIS

PLANTS AT OMAHA-ST. LOUIS-NEW ORLEANS

Falstaff Brewing Co.
Cranston, RI | 1970s/1980s

JOHN ADAMS
1797-1801 FEDERALIST

"No taxation without representation"

Falstaff Brewing Co.
Cranston, RI | 1970s/1980s

Fehr's XL

Short for "Excellent Lager," Fehr's XL was the flagship brand for the Louisville-based company. In the 1880s, the beer received gold medals at Louisville's Southern Exposition.

By 1901, the Frank Fehr Brewing Company was the largest brewery in Louisville. Fehr's was the only brewery of the Central Consumer's Company to reopen after Prohibition, but it was later sold to Schoenling in the late 1960s.

Frank Fehr Brewing Co., Inc.
Louisville, KY | 1940s/1950s

Frank Fehr Brewing Co., Inc.
Louisville, KY | 1940s/1950s

Feldschlösschen Brewery
Rheinfelden, Switzerland | 1970s

Genesee Brewing Co., Inc.
San Francisco, CA | 1970s/1980s

Fitger Brewing Co.
Duluth, MN | 1930s/1940s

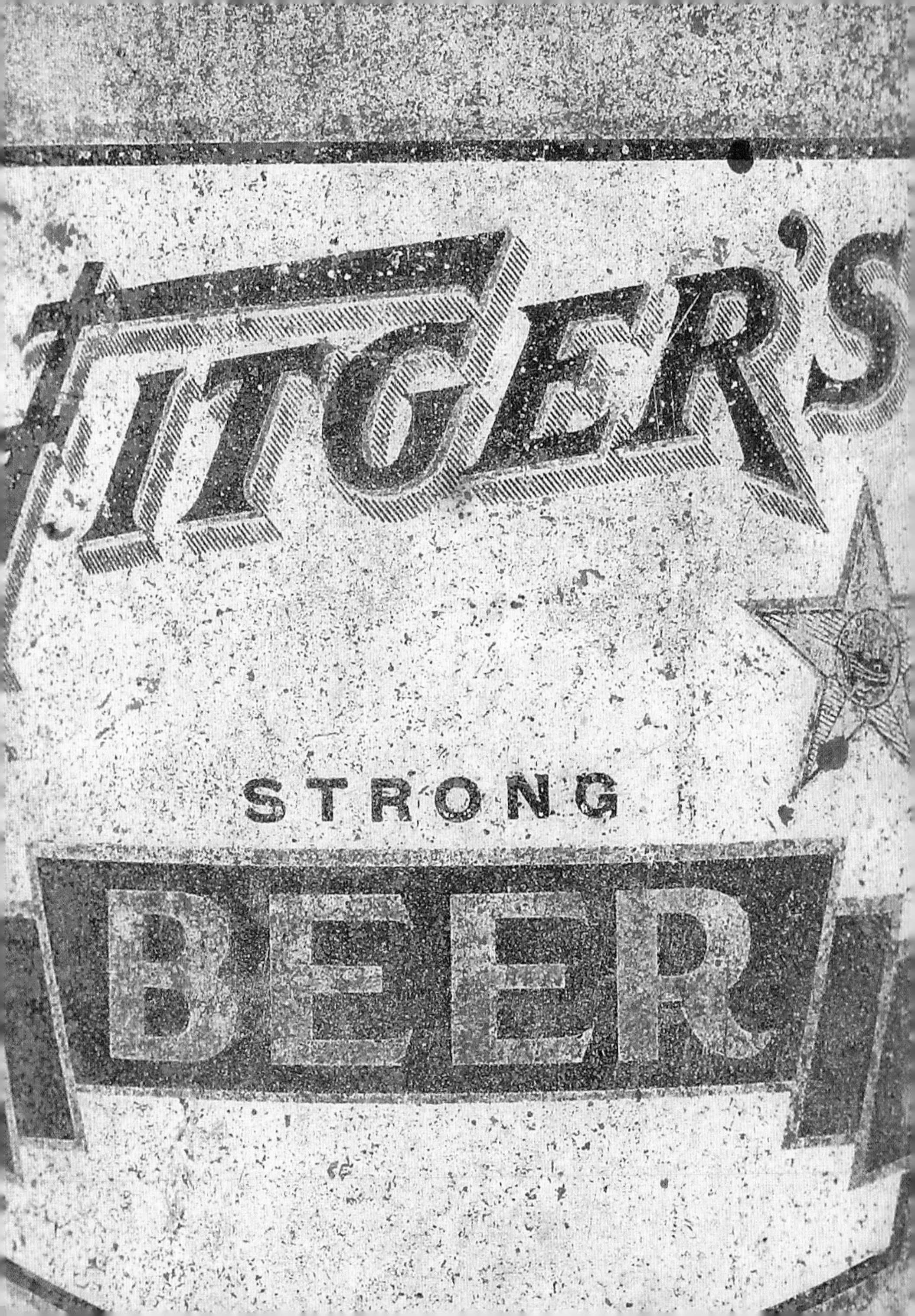
STRONG
BEER

Fitzgerald Bros. Brewing Co.
Troy, NY | 1940s/1950s

Fitzgerald Bros. Brewing Co.
Troy, NY | 1940s/1950s

Hellenic Breweries
Athens, Greece | 1970s

Fort Schuyler Brewing Co.
Utica, NY | 1970s/1980s

Fountain Brewing Co.
Fountain City, WI | 1940s/1950s

Heilman Brewing Co.
La Crosse, WI | 1960s

Frankenmuth Brewing Co.
Frankenmuth, MI | 1940s

Frankenmuth Brewing Co.
Frankenmuth, MI | 1950s

Frydenlunds Brewery
Oslo, Norway | 1970s

The Genesee Brewing Co.
Rochester, NY | 1970s/1980s

Gablinger's

Named after Swiss chemist Hersch Gablinger,
Gablinger's beer was the world's first light beer.
Poor marketing as a "diet beer" led to a steep
decline in sales, eventually leading to the failure
of the company. The recipe was eventually
acquired by Miller Brewing, where it is now
sold as Miller Lite.

Forrest Brewing Co.
New Bedford, MA | 1970s

Grace Bros. Brewing Co.
Santa Rosa, CA | 1930s

Grace Bros. Brewing Co.
Santa Rosa, CA | 1940s/1950s

Generic

Many generic beers were produced for stores around the country as cheaper alternatives to brand-name beer.

Forrest Brewing Co.
New York, NY | 1970s

Old Dutch Brewing Co.
Allentown, PA | 1970s

James Hanley Co.
New Orleans, LA | 1970s/1980s

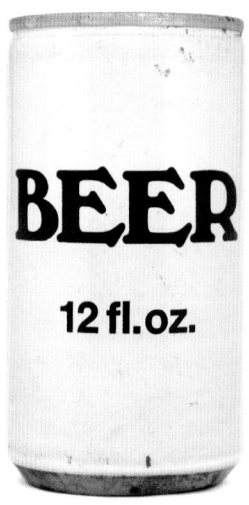

Du Bois Brewing Co.
Pittsburgh, PA | 1980s

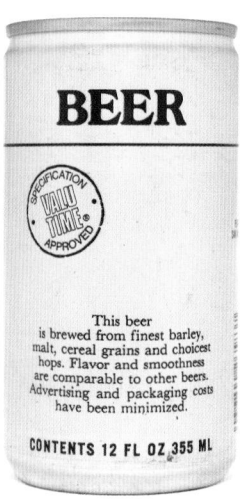

Pearl Brewing Co.
San Antonio, TX | 1980s

Falstaff Brewing Co.
Cranston, RI | 1980s

Falstaff Brewing Co.
Cranston, RI | 1980s

Pocono Brewing Co.
Wilkes-Barre, PA | 1980s

Falstaff Brewing Co.
Fort Wayne, IN | 1980s

Genesee Brewing Co.
Rochester, NY | 1970s/1980s

Genesee Brewing Co., Inc.
Rochester, NY | 1980s/1990s

A. Gettelman Brewing Co.
Milwaukee, WI | 1940s/1950s

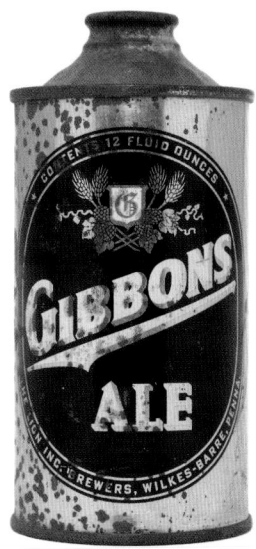

The Lion Brewery, Inc.
Wilkes-Barre, PA | 1930s/1940s

Gipps Brewing Co.
Peoria, IL | 1950s

The Glasgo Brewing Co., Inc.
Norfolk, VA | 1940s/1950s

PILSENER
PALE

Gluek

G

BE

Gluek's

In 1857, German immigrant and former lumberjack Gottlieb Gluek founded the Mississippi Brewing Company in Minneapolis, Minnesota. The name of the brewery later changed to the Gluek Brewing Company after its founder. Before the advent of refrigeration, Gluek's beer was lagered in caves on Nicollet Island in the Mississippi River.

Gluek Brewing Co.
Minneapolis, MN | 1930s/1940s

Gluek Brewing Co.
Minneapolis, MN | 1940s/1950s

Gluek Brewing Co.
Minneapolis, MN | 1940s/1950s

Gluek Brewing Co.
Minneapolis, MN | 1950s

Goebel Brewing Co.
Oakland, CA | 1940s

Goebel Brewing Co.
Detroit, MI | 1940s/1950s

M. K. Goetz Brewing Co.
St. Joseph, MO | 1940s/1950s

M. K. Goetz Brewing Co.
St. Joseph, MO | 1940s/1950s

Pearl Brewing Co.
San Antonio, TX | 1980s

El Rey Brewing Co.
San Francisco, CA | 1930s

Grace Bros. Brewing Co.
Santa Rosa, CA | 1940s

CONTENTS 12 FLUID OUNCES

AS GOOD AS GOLD FROM BREWERS OF OLD

Gold Seal
BEER

INTERNAL REVENUE TAX PAID
Does not contain more than 4% Alcohol by weight

MUTUAL BREWING CO. INC.
Ellensburg, Washington

The Mutual Brewing Co., Inc.
Ellensburg, WA | 1940s

Golden Age Breweries
Spokane, WA | 1930s/1940s

Grace Bros. Brewing Co.
Santa Rosa, CA | 1950s

Maier Brewing Co.
Los Angeles, CA | 1950s

Gotham Fine

Formed from Delatron Brewing in 1946, the Cincinnati Brewing Company marketed this beer as "jetter controlled," claiming that their unique canning process made Gotham beer taste better than others.

Cincinnati Brewing Co.
Cincinnati, OH | 1940s/1950s

Minneapolis Brewing Co.
Minneapolis, MN | 1930s/1940s

Minneapolis Brewing Co.
Minneapolis, MN | 1940s/1950s

Minneapolis Brewing Co.
Minneapolis, MN | 1950s

Grain Belt Breweries Inc.
Minneapolis, MN | 1970s

STRONG

Premium

Grain Belt

The Friendly Beer

MINNEAPOLIS BREWING CO.
MINNEAPOLIS, MINN.

GREAT AMERICAN BEER

NET CONTENTS 12 FL. OZ.

General Brewing Co.
Vancouver, WA | 1980s

Great Falls Breweries Inc.
Great Falls, MT | 1950s

Grolsch Brewery
Enschede, The Netherlands | 1980s

Toohey's Ltd.
Sydney, Australia | 1990s

Arthur Guinness & Son
Dublin, Ireland | 1990s

Gunther Brewing Co.
Baltimore, MD | 1950s

Gunther Brewing Co.
Baltimore, MD | 1970s/1980s

Joseph Huber Brewing Co.
Monroe, WI | 1980s

A. Haas Brewing Co.
Houghton, MI | 1940s

D.G. Yuengling & Son Inc.
Pottsville, PA | 1970s

Narragansett Brewing Co.
Cranston, RI | 1970s/1980s

Theodore Hamm Brewing Co.
San Francisco, CA | 1950s

Theodore Hamm Brewing Co.
St. Paul, MN | 1950s

Hamm

DRAFT BEER

12 FLUID OZS.

Draft brewed from choicest barley malt,

Hamm's Draft

This keg-shaped can from the 1970s proved to be a popular trend in beer can design, due to its departure from the traditional can shape.

Theodore Hamm Co.
St. Paul, MN | 1970s

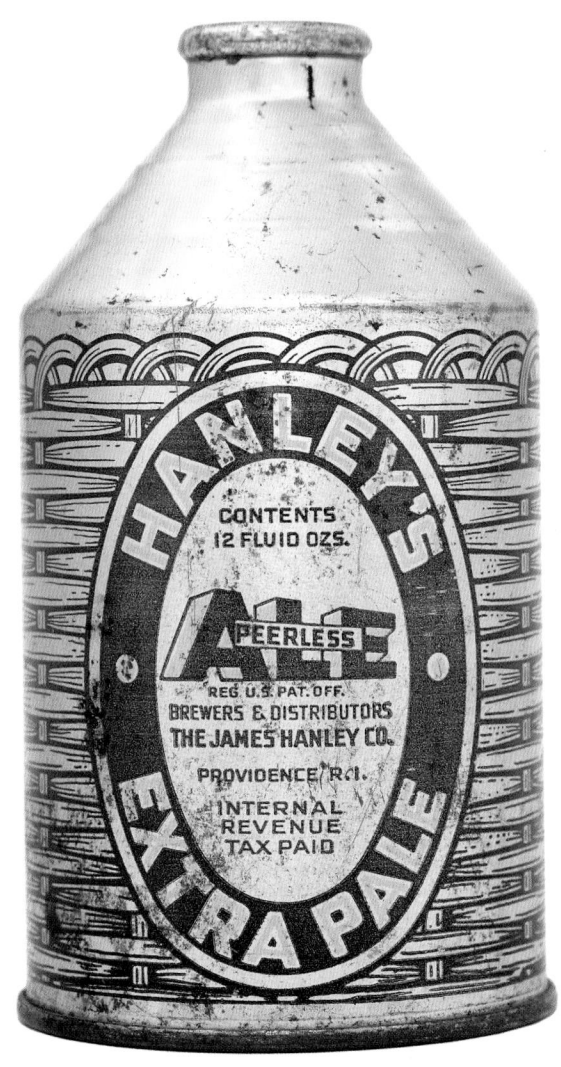

The James Hanley Co.
Providence, RI | 1940s/1950s

The James Hanley Co.
Cranston, RI | 1970s

Joseph Huber Brewing Co.
Monroe, WI | 1970s

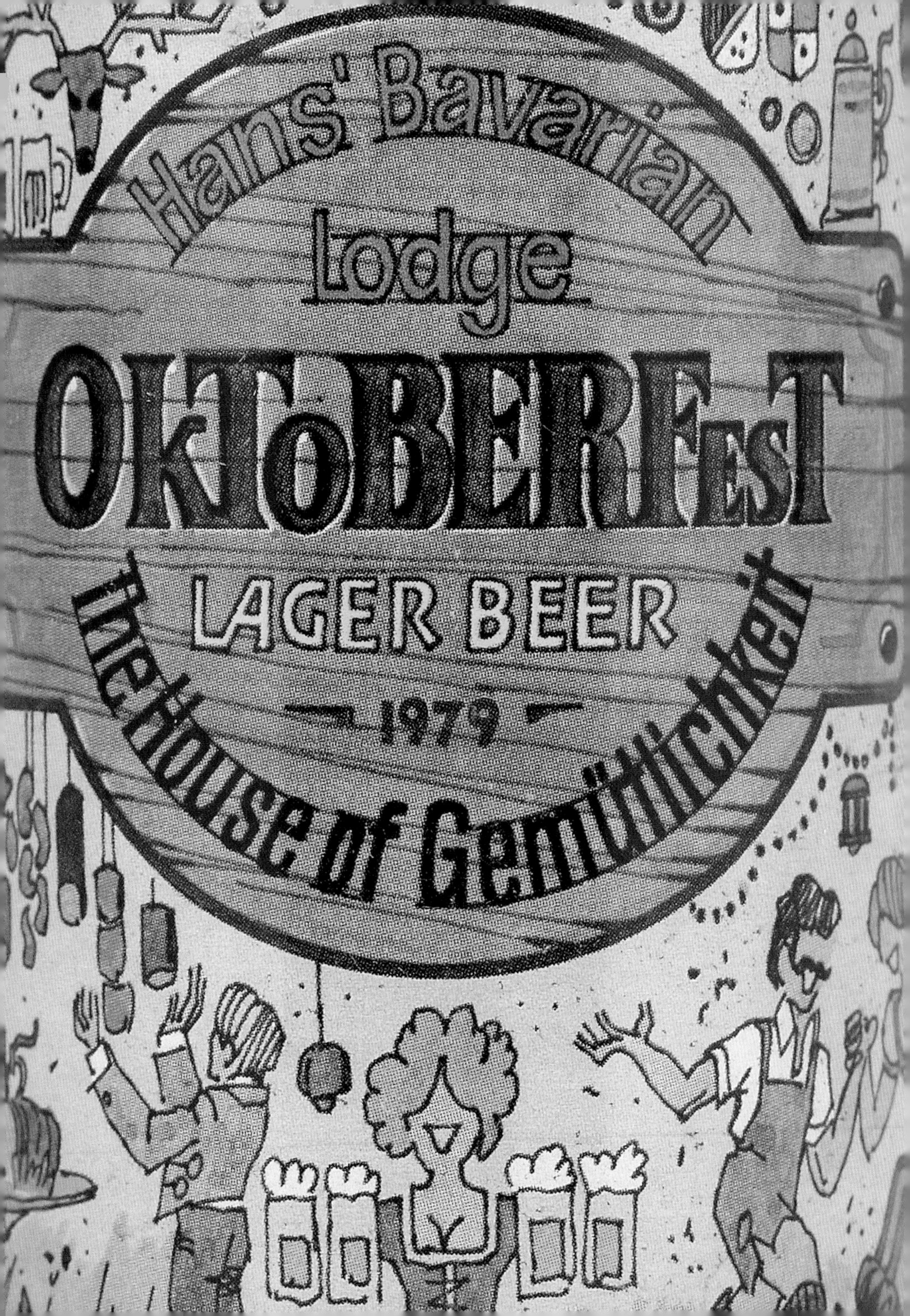
Hans' Bavarian
Lodge
OKTOBERFEST
LAGER BEER
1979
The House of Gemütlichkeit

Walter Brewing Co.
Eau Claire, WI | 1970s/1980s | front

Walter Brewing Co.
Eau Claire, WI | 1970s/1980s | back

Joseph Huber Brewing Co.
Monroe, WI | 1980s

Harvard Brewing Co.
Lowell, MA | 1950s

John Hauenstein Co.
New Ulm, MN | 1950s

Hedrick Brewing Co.
Hammonton, NJ | 1970s

Heidel Bräu Brewing Co.
La Crosse, WI | 1980s

Heidelberg Brewing Co.
Tacoma, WA | 1960s

Heineken Brouwerijen B.V.
Amsterdam, The Netherlands | 1970s

Joseph Huber Brewing Co.
Monroe, WI | 1970s

Missoula Brewing Co.
Missoula, MT | 1940s

NONE FINER—ANYWHERE

Highlander

PREMIUM
BEER

MISSOULA, MONTANA · CONTENTS 12 FLUID OUNCES

Missoula Brewing Co.
Missoula, MT | 1950s

Rainier Brewing Co.
Seattle, WA | 1970s/1980s

Brauerei Abfüllung
Munich, Germany | 1970s/1980s

Hoff-Brau Brewing Co.
Fort Wayne, IN | 1940s/1950s

Horlacher Brewing Co.
Allentown, PA | 1970s

The Hudepohl Brewing Co.
Cincinnati, OH | 1940s/1950s

The Hudepohl Brewing Co.
Cincinnati, OH | 1940s/1950s

The Hudepohl Brewing Co.
Cincinnati, OH | 1950s

The Hudepohl Brewing Co.
Cincinnati, OH | 1950s

Hudepohl Brewing Co.
Cincinnati, OH | 1960s

Hudepohl Brewing Co.
Cincinnati, OH | 1970s

Hu-Dey

This commemorative can by the Hudepohl Brewing Company paid homage to the Cincinnati Bengals football team and featured their chant: "Who Dey Think Gonna Beat Dem Bengals?"

The Hudepohl Brewing Co.
Cincinnati, OH | 1980s

Hümmer-Bräu
Dingolshausen, Germany | 1970s/1980s

A. Hürlimann Brauerei A.G.
Zurich, Switzerland | 1970s/1980s

Cerveceria India Inc.
Mayagüez, Puerto Rico | 1970s/1980s

F. & M. Schaefer Brewing Co.
Albany, NY | 1950s

Pittsburgh Brewing Co.
Pittsburgh, PA | 1940s/1950s

Pittsburgh Brewing Co.
Pittsburgh, PA | 1950s

Pittsburgh Brewing Co.
Pittsburgh, PA | 1970s

Pittsburgh Brewing Co.
Pittsburgh, PA | 1970s/1980s

Iroquois Beverage Corp.
Buffalo, NY | 1940s/1950s

Iroquois Beverage Corp.
Buffalo, NY | 1960s

Jax

This 1974 Jax beer can features an illustration of the famous Andrew Jackson statue in Jackson Square, New Orleans. The depiction of the first equestrian statue in the United States became the signature trademark for the Jax brand and evolved with greater levels of detail in the following years.

Jackson Brewing Co.
New Orleans, LA | 1970s

Duluth Brewing and Malting Co.
Duluth, MN | 1960s

Karlsberg Brauerei K.G.
Homburg, Germany | 1970s

Mankato Brewing Co.
Mankato, MN | 1940s/1950s

Maier Brewing Co.
Los Angeles, CA | 1950s

Cold Spring Brewing Co.
Cold Spring, MN | 1970s

Kessler Brewing Co.
Helena, MT | 1940s/1950s

Kingsbury Breweries Co.
Sheboygan, WI | 1940s/1950s

Kingsbury Breweries Co.
Sheboygan, WI | 1950s

Jacob Ruppert Brewery
Orange, NJ | 1970s/1980s

C. Schmidt's & Sons Brewery
Philadelphia, PA | 1970s/1980s

Erie Brewing Co.
Erie, PA | 1940s/1950s

Erie Brewing Co.
Erie, PA | 1970s/1980s

REG. U.S. PA... ...E.

Krueger Cream Ale

In 1934, the Gottfried Krueger Brewing Company became the first brewery to can beer. Based in Newark, New Jersey, Krueger tested the concept of canned beer in Richmond, Virginia, isolated from its core demographic in the event that the concept flopped.

Obviously, the idea was far from a failure. While the first cans ever made were flat-tops like the one below, Krueger went on to sell beer in cone-top cans as well.

G. Krueger Brewing Co.
Newark, NJ | 1930s

Krueger Brewing Co.
Wilmington, DE | 1940s

Rainier Brewing Co.
San Francisco, CA | 1930s/1940s

Kuebler Brewing Co Inc.
Easton, PA | 1950s

Labatt Breweries Ltd.
Montreal, Canada | 1970s

Labatt Breweries Ltd.
London, Canada | 1970s

Lamot Brewery
Mechelen, Belgium | 1970s

Jacob Leinenkugel Brewing Co.
Chippawa Falls, WI | 1940s/1950s

Jacob Leinenkugel Brewing Co.
Chippawa Falls, WI | 1970s

The Leisy Brewing Co.
Cleveland, OH | 1940s/1950s

Leopard Brewery Ltd.
Hastings, New Zealand | 1970s

Lion

With a logo clearly resembling its name, this Lion Pilsener beer was a product of Pilsener Brewing in New York—one of several companies that had a relationship with the brand's original brewer: Lion Brewing.

Pilsener Brewing Co.
New York, NY | 1940s/1950s

Lion Breweries Ltd.
Christchurch, New Zealand | 1970s

The Lion Brewery, Inc.
Wilkes-Barre, PA | 1970s/1980s

Rainier Brewing Co.
San Francisco, CA | 1930s

Jacob Leinenkugel Brewing Co.
Chippewa Falls, WI | 1970s

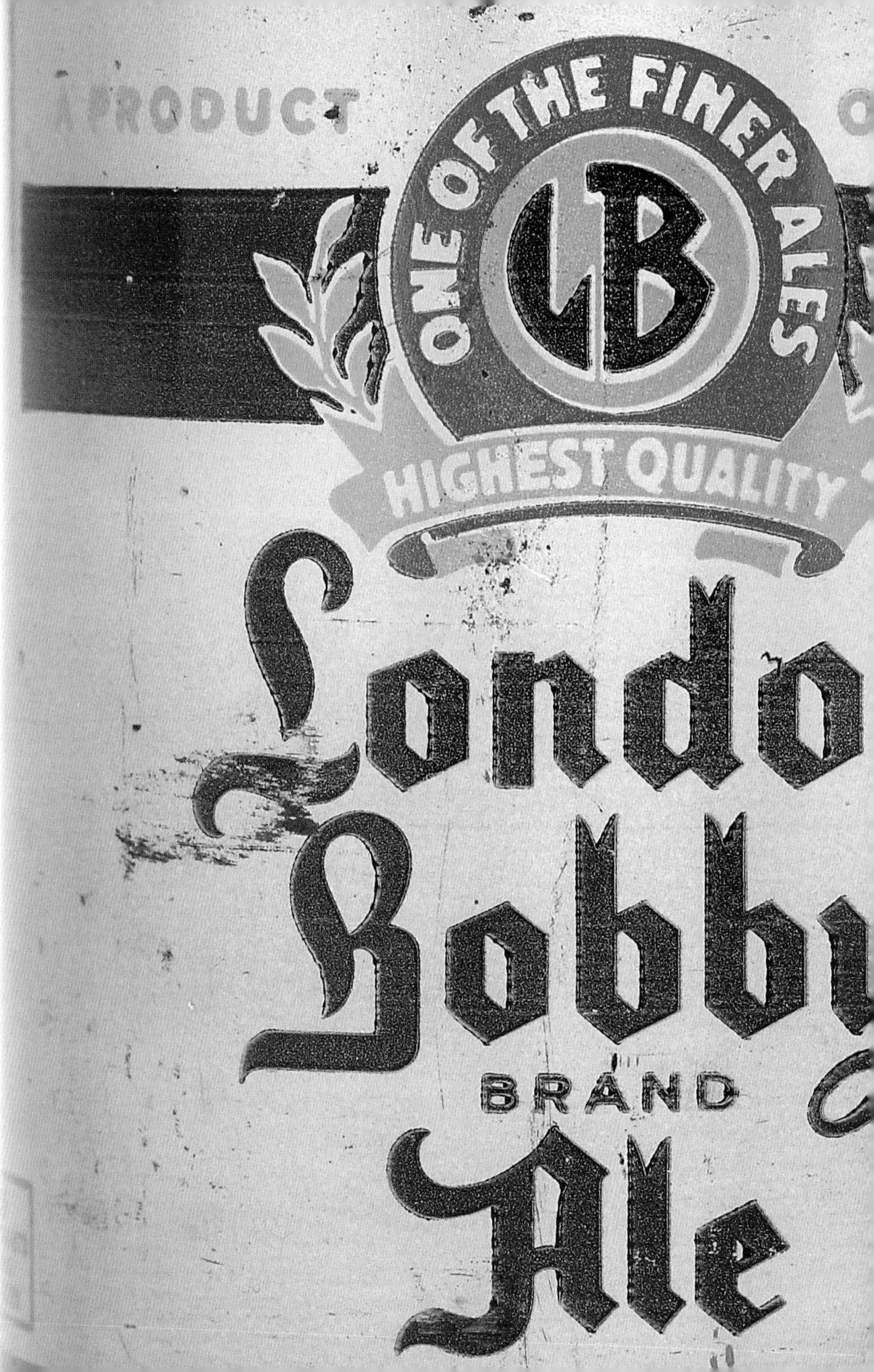
PRODUCT
ONE OF THE FINER ALES
LB
HIGHEST QUALITY
BRAND
Ale

London Bobby

With metal in lower demand in a post–World War II era, the Miami Valley Brewing Company of Dayton, Ohio, began using cans as their business grew and their annual capacity expanded.

These uniquely shaped crowntainer cans could be filled on a standard bottling line with few adjustments. Such versatility attracted small breweries that couldn't afford an additional production line for flat-top cans.

The Miami Valley Brewing Co.
Dayton, OH | 1940s/1950s

El Dorado Brewing Co.
Stockton, CA | 1930s/1940s

London Tavern

EL DORADO BREWING CO.

Lone Star Brewing Co.
San Antonio, TX | 1970s/1980s

Löwenbräu A.G.
Munich, Germany | 1970s/1980s

AGE
DATED
BEER

Lucky Lager

Artist and art director Charles Stafford Duncan of McCann-Erickson advertising in San Francisco designed this label for General Brewing in the 1930s. Duncan's design became an iconic symbol of the brand that stood out among its competitors.

Lucky Lager became so successful that its creator, General Brewing, changed its name to Lucky Lager Brewing in 1948.

Lucky Lager Brewing Co.
Vancouver, WA | 1940s/1950s

Scottish & Newcastle Breweries Ltd.
Edinburgh, Scotland | 1970s

Carling O'Keefe Breweries
Toronto, Canada | 1970s

Maier Brewing Co.
Los Angeles, CA | 1930s/1940s

Maier Brewing Co.
Los Angeles, CA | 1940s

Maier Brewing Co.
Los Angeles, CA| 1950s

Yakima Valley Brewing Co.
Selah, WA| 1940s

Genuine

MARTIN'S

BEER

Peter Hand Brewery Co.
Chicago, IL | 1950s

Miller Brewing Co.
Milwaukee, WI | 1980s/1990s

Meister Bräu Lite

This Lite beer originated as Gablinger's Diet Beer, the recipe for which was given to Meister Bräu in 1967. Miller Brewing later acquired Mesiter Bräu in 1972 and adopted the "Lite" trademark that adorns this can.

Miller Brewing Co.
Milwaukee, WI | 1970s/1980s

Menominee-Marinette Brewing Co.
Menominee, MI | 1940s/1950s

CONTENTS 12
OZ.
MENOMINEE-MARINE
TRADE MARK
CT BEER
Menominee
CHAMPION
"THE BEST
WHAT IS"
LIGHT
BEER

CONTENTS 12 FL. OZ.

SINCE 1864

Metz

JUBILEE   BEER

METZ BREWING COMPANY
OMAHA, NEBRASKA

Metz Brothers Brewing Co.
Omaha, NE | 1950s

Anheuser-Busch Inc.
St. Louis, MO | 1970s

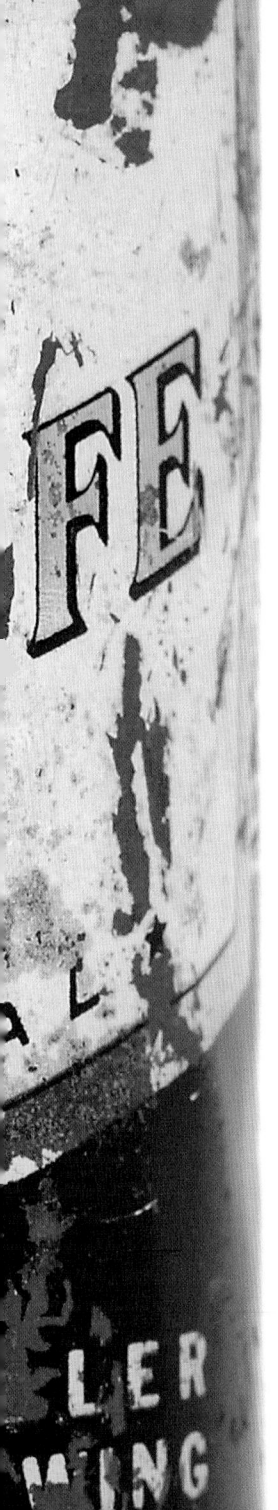

Miller

Founded in 1855 by Fredrick Miller in Milwaukee, Wisconsin, the Miller Brewing Company grew to be one of the largest brewers in the United States.

Throughout its lifespan, Miller has created a number of iconic brands among beer drinkers, including its oldest brand, Miller High Life.

Miller Brewing Co.
Milwaukee, WI | 1940s/1950s

Miller Brewing Co.
Milwaukee, WI | 1950s | front

Miller Brewing Co.
Milwaukee, WI | 1950s | back

Miller Brewing Co.
Milwaukee, WI | 1960s

Miller Brewing Co.
Milwaukee, WI | 1970s/1980s

O'Keefe Ltd.
Montreal, Canada | 1980s/1990s

Waukee Brewing Co.
Hammonton, NJ | 1970s

Joseph Schlitz Brewing Co.
Milwaukee, WI | 1930s

Joseph Schlitz Brewing Co.
Milwaukee, WI | 1930s/1940s

MOLSO

CANADI

BIÈRE *Lager*

Molson

This Canadian brewery was founded in 1786 in Montreal, Canada, making it the oldest brewery in North America. A range of icons have been associated with the brand and its variants, most notably the maple leaf from Canada's flag, declaring the brand's patriotism.

Molson Breweries
Montreal, Canada | 1970s

Molson Breweries
Montreal, Canada | 1970s

Molson Breweries
Montreal, Canada | 1980s/1990s

Moosehead Breweries Ltd.
Dartmouth, Canada | 1970s

Moretti S.P.A.
Udine, Italy | 1970s

Royal Brewing Co.
New Orleans, LA | 1970s

Geo. Muehlebach Brewing Co.
Kansas City, MO| 1950s

The Burger Brewing Co.
Akron, OH | 1950s

Pittsburgh Brewing Co.
Pittsburgh, PA | 1970s

Brasserie de Mützig
Mützig, France | 1970s/1980s

Narragansett Brewery
Cranston, RI | 1960s

Narragansett Brewery
Cranston, RI | 1970s/1980s

National Bohemian

Often referred to as "Natty Boh," National Bohemian was the creation of National Brewing in Baltimore, Maryland, where it became the "official" beer of the city.

The brand's mascot, Mr. Boh—a man with one eye and a handlebar mustache—evolved with the brand over the years. Mr. Boh remains a recognizable icon of the brand to this day.

National Brewing
Baltimore, MD | 1950s

Carling National Breweries
Baltimore, MD | 1970s/1980s

National Brewing
Baltimore, MD | 1950s

National Brewing
Baltimore, MD | 1970s/1980s

Louis F. Neuweiler's Sons
Allentown, PA | 1940s

ENTS
OZS.

INT. REVEN
TAX PAI

NATIONAL

PALE · DRY · BEER

Newcastle Breweries Ltd.
Newcastle, England | 2000s

Washington Breweries Inc.
Columbus, OH | 1940s/1950s

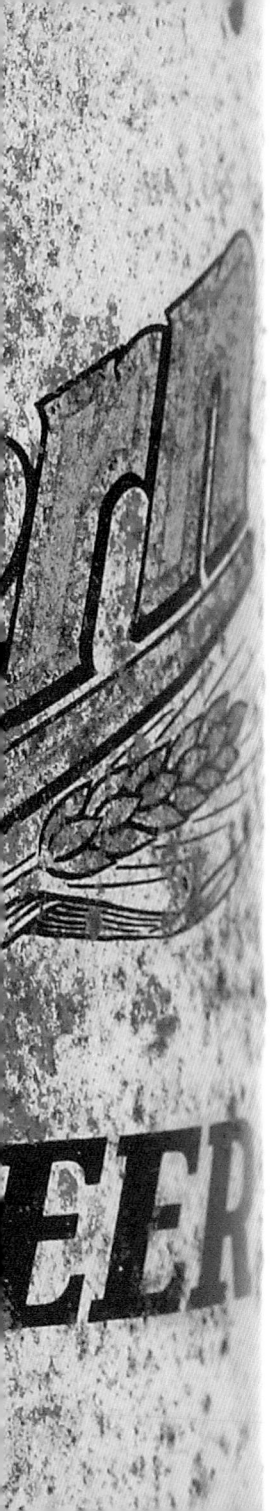

Northern

Located in the northern reaches of Wisconsin, Northern Brewing became one of the most successful industries in the region. However, in the 1960s, a bad batch of beer resulted in returns and the loss of longtime accounts. In 1967, the company ceased production, and was eventually bought by Cold Spring Brewing of Minnesota.

Northern Brewing Co.
Superior, WI | 1950s

August Schell Brewing Co.
New Ulm, MN | 1970s

Oconto Brewing Co.
Oconto, WI | 1950s/1960s

Oertel Brewing Co., Inc.
Louisville, KY | 1940s/1950s

Molson Breweries
Etobicoke, Canada | 1970s

Eastern Brewing Co.
Hammonton, NJ | 1970s

Peter Hand Brewing Co.
Chicago, IL | 1970s

Peter Hand Brewing Co.
Chicago, IL | 1970s

Maier Brewing Co.
Los Angeles, CA | 1950s

Pittsburgh Brewing Co.
Pittsburgh, PA | 1970s

The Cumberland Brewing Co.
Cumberland, MD | 1940s/1950s

Northampton Brewing Co.
Northampton, PA | 1940s/1950s

Old Georgetown

Initially only available on draft, Old Georgetown beer was first canned in the spring of 1950 by the Christian Heurich Brewery of Washington, D.C. The beer was fairly successful, and became the brewery's centerpiece in its marketing materials.

Multiple versions of this can exist, each featuring a map of the United States capital, highlighting some of the city's many landmarks.

Christian Heurich Brewing Co.
Washington, D.C. | 1940s

THE PRIDE OF WISCONSIN

Old Imperial BEER

Rahr-Green Bay Brewing Co.
Green Bay, WI | 1940s/1950s

Standard Brewing Co., Inc.
Rochester, NY | 1940s/1950s

The Old Reading Brewery Inc.
Reading, PA | 1930s

Old Reading Brewery Inc.
Reading, PA | 1940s/1950s

G. Heileman Brewing Co.
La Crosse, WI | 1960s/1970s

G. Heileman Brewing Co.
La Crosse, WI | 1930s/1940s

G. Heileman Brewing Co.
La Crosse, WI | 1950s

Pabst Brewing Co.
Los Angeles, CA | 1970s/1980s

Walter Brewing Co.
Eau Claire, WI | 1970s

Rochester Brewing Co.
Rochester, NY | 1940s/1950s

Carling O'Keefe Breweries
Toronto, Canada | 1970s

Olde Frothingslosh

Originally created as a joke by American radio personality Regis Cordic, Olde Frothingslosh was created by a local Pittsburgh brewer as a seasonal promotion. The beer became so popular that over thirty different cans were designed and produced.

Carrying the tagline "The Pale Stale Ale with the Foam on the Bottom," these joke cans continue to appeal to collectors today.

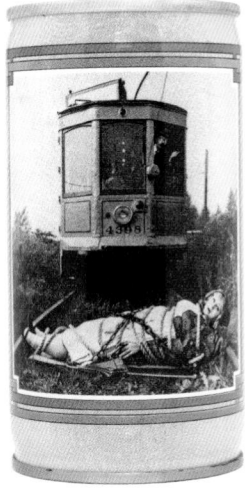

Pittsburgh Brewing Co.
Pittsburgh, PA | 1970s | front & back

Pittsburgh Brewing Co.
Pittsburgh, PA | 1970s/1980s

Pittsburgh Brewing Co.
Pittsburgh, PA | 1980s

Pittsburgh Brewing Co.
Pittsburgh, PA | 1980s

Pittsburgh Brewing Co.
Pittsburgh, PA | 1980s

The Erie Brewing Co.
Erie, PA | 1970s

Maier Brewing Co.
Los Angeles, CA | 1950s

The New Philadelphia Brewing Co.
New Philadelphia, OH | 1940s/1950s

Olympia Brewing Co.
Olympia, WA | 1950s

Oranjeboom Breweries
Rotterdam, The Netherlands | 1970s/1980s

Orion Breweries Ltd.
Nago, Japan | 1970s

Orion Breweries Ltd.

Nago, Japan | 2000s

Pabst

In August 1935, Pabst became the second brewer to put beer in a can. The canned beer was initially labeled "Export" before the famous Pabst Blue Ribbon name was adopted and the iconic ribbon implemented into the can's label.

The name "Blue Ribbon" originated in 1882, when Pabst began tying blue silk ribbons to its bottled beer. Over 100 years later, Pabst Blue Ribbon, or PBR, remains one of the most iconic beer brands in history.

Pabst Brewing Co.
Milwaukee, WI | 1930s

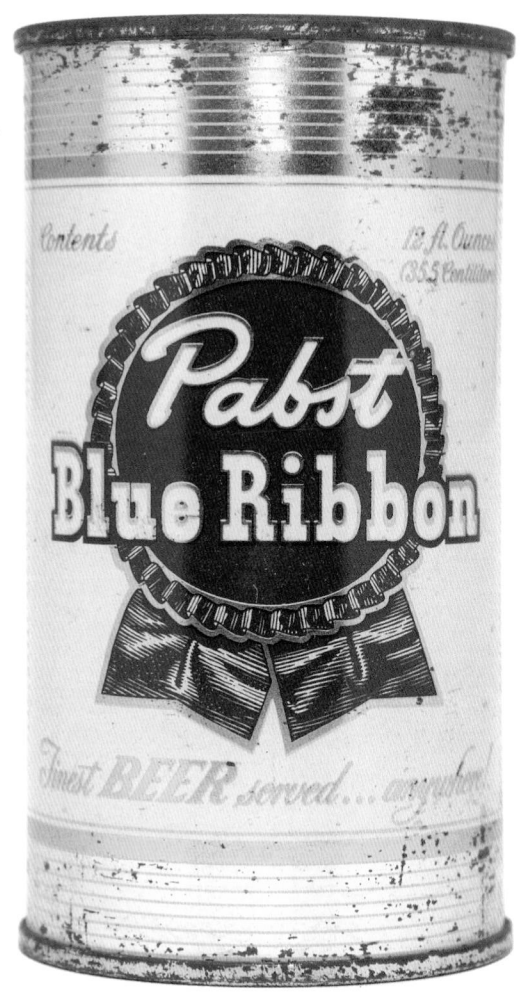

Pabst Brewing Co.
Milwaukee, WI | 1950s

Pabst Brewing Co.
Milwaukee, WI | 1960s

Rainier Brewing Co.
San Francisco, CA | 1930s

Maier Brewing Co.
Los Angeles, CA | 1960s

Patrizier Bräu A.G.
Nuremberg, Germany | 1970s

Paulaner-Salvator Thomasbrau A.G.
Munich, Germany | 1970s

Peerless BEER

BREWED AND FILLED BY
LA CROSSE BREWERIES INC., LA CROSSE, WIS.

La Crosse Breweries Inc.
La Crosse, WI | 1940s/1950s

Lebanon Valley Brewing Co.
Lebanon, PA | 1960s

Birra Peroni
Rome, Italy | 1970s

Joseph Huber Brewing Co.
Monroe, WI | 1970s

Piels Bros. Brewing Co.
Brooklyn, NY | 1960s

Piels Bros.
Willimansett, MA | 1970s/1980s

Pilser Brewing Co.
Bronx, NY | 1930s

Metropolis Brewing Co.
New York, NY | 1960s

D.G. Yuengling & Son Inc.
Pottsville, PA | 1970s

Stevens Point Brewery
Stevens Point, WI | 1970s

Polar Brewing Co.
Hammonton, NJ | 1970s

Atlas Brewing Co.
Chicago, IL | 1940s

Gottlieb Heileman Brewing Co.
La Crosse, WI | 1970s/1980s

Walter Brewing Co.
Eau Claire, WI | 1970s

August Schell Brewing Co.
New Ulm, MN | 1970s

Rainier Brewing Co.
San Francisco, CA | 1930s

Rainier

The history of Rainier beer dates back to 1878 when it was first brewed in Seattle, Washington. In 1916, alcoholic beverages were outlawed in Washington State, and as a result, Rainier moved to San Francisco until Prohibition ended.

Throughout its history, Rainier was produced by several different breweries. The brand survives to this day under the ownership of Pabst.

Rainier Brewing Co.
San Francisco, CA | 1940s

Rainier Brewing Co.
San Francisco, CA | 1940s

Reading Brewing Co.
Philadelphia, PA | 1970s

Reading Brewing Co.
Reading, PA | 1970s

Reading Brewing Co.
Reading, PA | 1970s

Brewing Corp. of America
Cleveland, OH | 1940s/1950s

Carling Brewery
Cleveland, OH | 1950s

Carling Brewing Co.
Cleveland, OH | 1950s

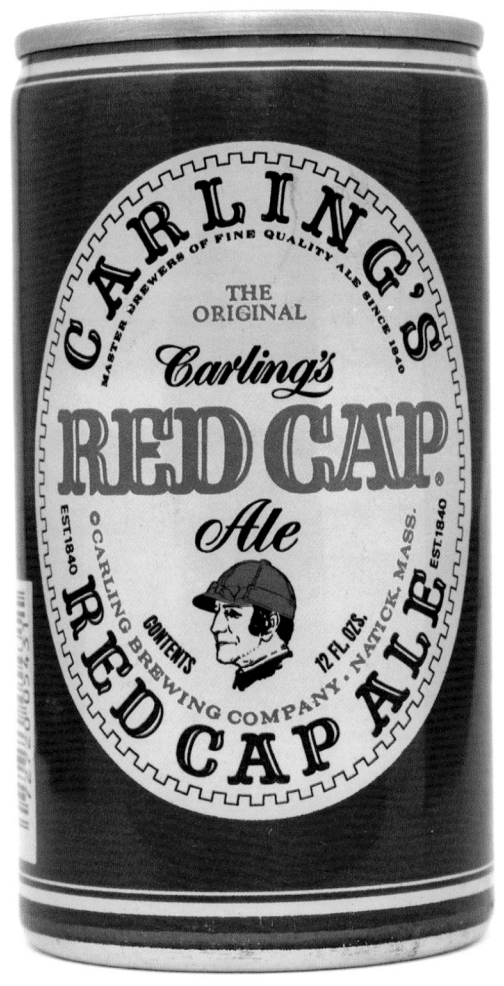

Carling Brewing Co.
Natick, MA | 1970s/1980s

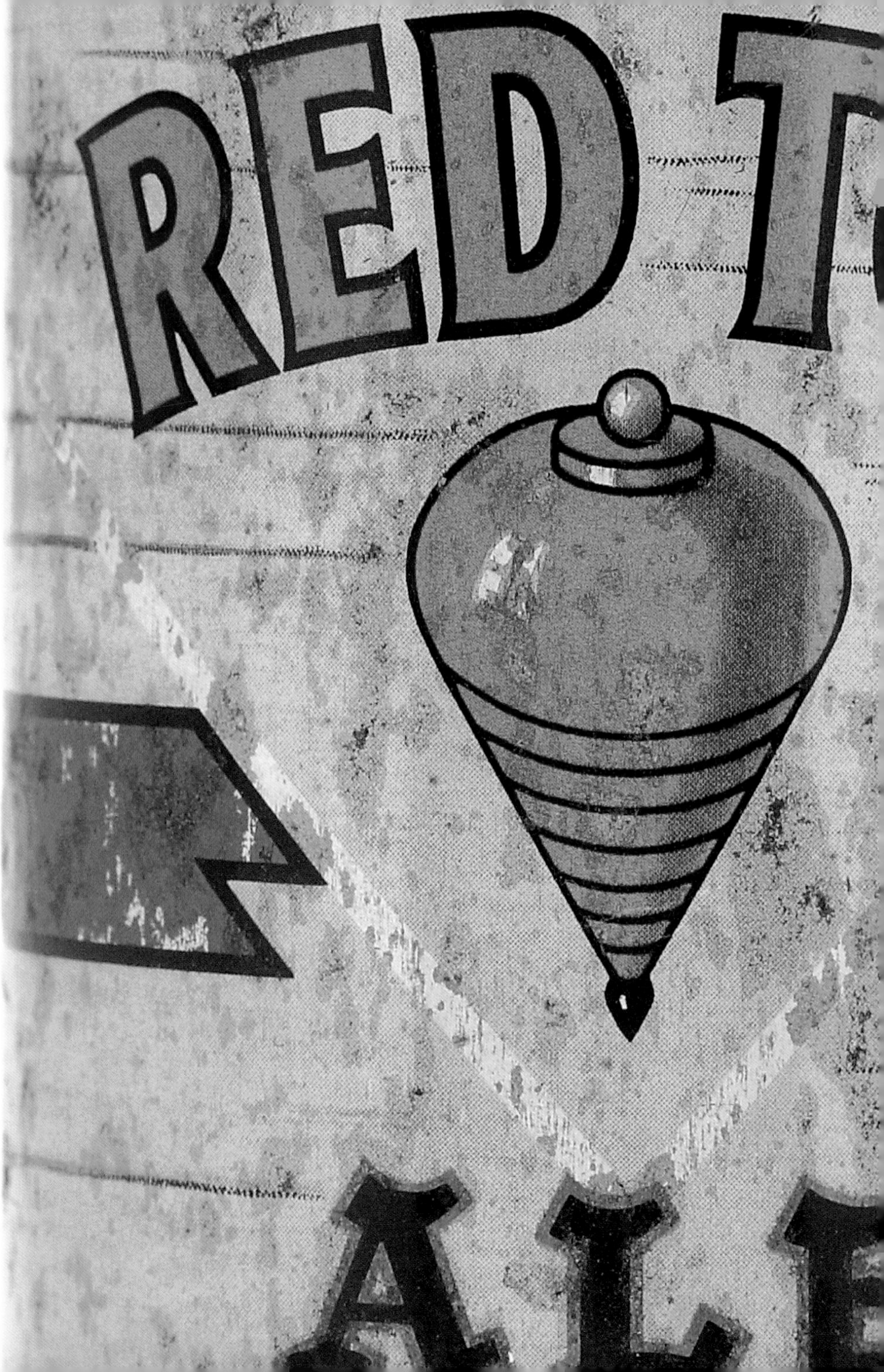
RED

Red Top Ale

The 1940s and early 1950s marked the height of success for Red Top Brewing. In 1954, the company was led by former New York Yankee Tommy Henrich, who headed it until 1956. A year later, after production problems and years of pressure from national brands, the final batch of Red Top was produced in Cincinnati.

Red Top Brewing Co.
Cincinnati, OH | 1940s/1950s

Red Top Brewing Co.
Cincinnati, OH | 1950s

Red Top Brewing Co.
Cincinnati, OH | 1950s

Regal Amber Brewing Co.
San Francisco, CA | 1940s

The People's Brewing Co.
Duluth, MN | 1950s

General Brewing Co.
San Francisco, CA | 1970s

Reisch Brewing Co.
Springfield, IL | 1940s/1950s

The Renner Co.
Youngstown, OH | 1940s/1950s

Liebmann Breweries Inc.
New York, NY | 1950s

Richbrau

Despite its looks, this Richbrau can, manufactured by the Home Brewing Company, was certainly not one of the first beer cans ever made. In fact, Home Brewing didn't install a canning line until 1952, some eighteen years after the first canned beers were produced.

Home Brewing Co., Inc.
Richmond, VA | 1950s

Pittsburgh Brewing Co.
Pittsburgh, PA | 1970s/1980s

Cream Ale

Dixie Brewing Co.
New Orleans, LA | 1980s | front

Dixie Brewing Co.
New Orleans, LA | 1980s | back

Anaconda Brewing Co.
Anaconda, MT | 1940s/1950s

The South African Breweries
Johannesburg, South Africa | 1970s/1980s

Maier Brewing Co.
Los Angeles, CA | 1950s

The George Wiedemann Brewing Co.
Newport, KY | 1950s

The George Wiedemann Brewing Co.
Newport, KY | 1950s

Duluth Brewing & Malting Co.
Duluth, MN | 1940s/1950s

N.V. de Posthoorn
Rotterdam, The Netherlands | 1970s

Koller Brewing Co.
Chicago, IL | 1930s/1940s

Rainier Brewing Co.
San Francisco, CA | 1930s

San Miguel Brewery Ltd.
Mandaluyong City, The Philippines | 1970s

Sapporo Breweries
Tokyo, Japan | 1970s/1980s

Walter Brewing Co.
Eau Claire, WI | 1980s

August Schell Brewing Co.
New Ulm, MN | 1940s/1950s

August Schell Brewing Co.
New Ulm, MN | 1970s

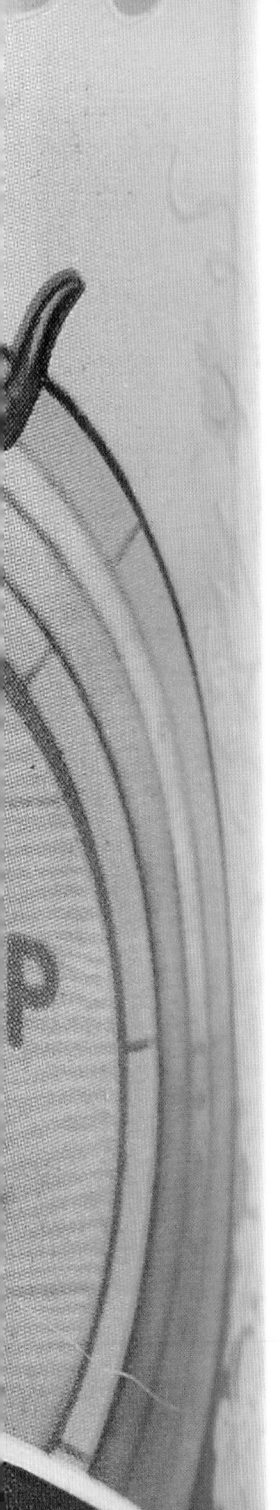

Schell's Bock

Goats and rams often made their way onto bock beer packaging, and this Schell's can is no exception. German for "billy goat," the term *bock* refers to beers that are typically strong, dark, and above average in alcohol content.

The phrase "Ein bock bier" or "One bock beer" gained popularity in Einbeck, Germany, and helped to establish the connection between the animal and the traditional winter brew.

August Schell Brewing Co.
New Ulm, MN | 1970s

August Schell Brewing Co.
New Ulm, MN | 1970s

August Schell Brewing Co.
New Ulm, MN | 1970s

August Schell Brewing Co.
New Ulm, MN | 1970s

August Schell Brewing Co.
New Ulm, MN | 1980s

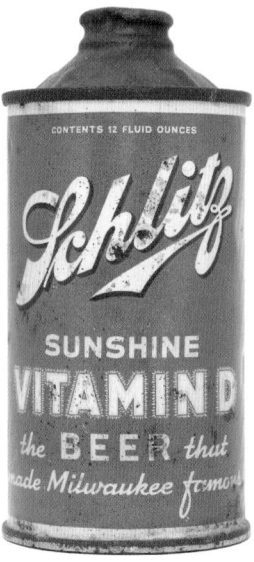

Joseph Schlitz Brewing Co.
Milwaukee, WI | 1930s

Joseph Schlitz Brewing Co.
Milwaukee, WI | 1940s/1950s

SCHELL'S

Xmas

Brew

Schlitz Brewing Co.
Milwaukee, WI | 1970s

Joseph Schilitz Brewing Co.
Milwaukee, WI | 1970s

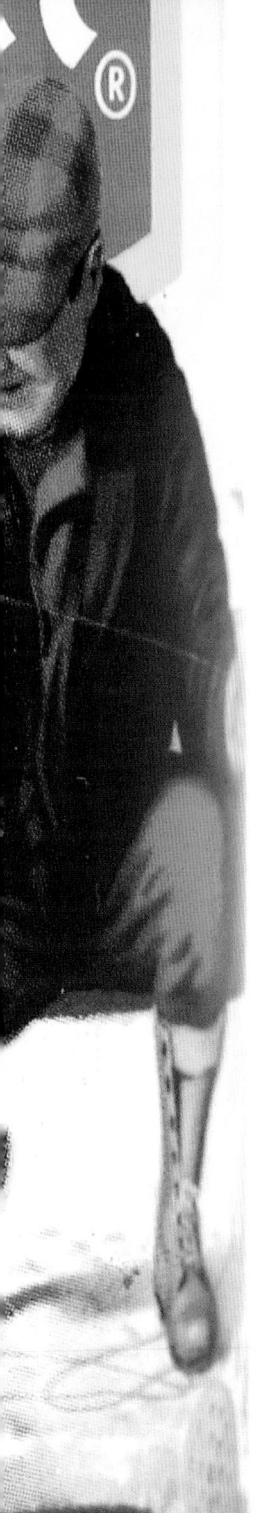

Schmidt

Often associated with the great outdoors, Schmidt beer appealed to many hunters and fishermen. Known as "animal beer," Schmidt cans, such as the one shown below, featured a range of animal and nature scenes. The brand's slogan, printed across the bottom of the cans, reads: "The Brew that grew with the Great Northwest."

Gottlieb Heileman Brewing Co.
La Crosse, WI | 1970s/1980s

Gottlieb Heileman Brewing Co.
La Crosse, WI | 1970s/1980s

Gottlieb Heileman Brewing Co.
La Crosse, WI | 1970s/1980s

Gottlieb Heileman Brewing Co.
La Crosse, WI | 1970s/1980s

Gottlieb Heileman Brewing Co.
La Crosse, WI | 1970s/1980s

Gottlieb Heileman Brewing Co.
La Crosse, WI | 1970s/1980s

The Schmidt Brewing Co.
Detroit, MI | 1940s/1950s

C. Schmidt & Sons Inc.
Philadelphia, PA | 1970s/1980s

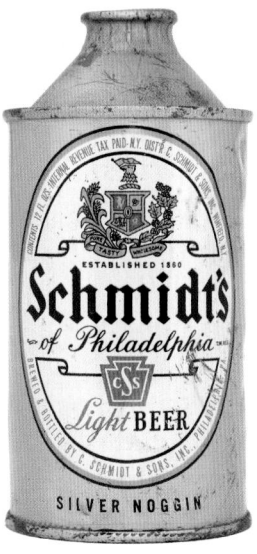

C. Schmidt & Sons Inc.
Philadelphia, PA | 1950s

C. Schmidt & Sons Inc.
Philadelphia, PA | 1960s/1970s

Jacob Schmidt Brewing Co.
St. Paul, MN | 1950s

C. Schmidt & Sons Inc.
Philadelphia, PA | 1950s

C. Schmidt's & Sons Brewery
Philadelphia, PA | 1970s/1980s

CONTENTS 12 FL. OZS.

CINCINNATI'S FINEST

Schoenling®

Lager

BEER

BREWED AND PACKAGED BY
SCHOENLING BREWING CO., CINCINNATI,

Schoenling Brewing Co.
Cincinnati, OH | 1950s

Schoenling Brewing Co.
Cincinnati, OH | 1970s

Seven Springs Mountain Beer

After Prohibition ended in the United States, many small breweries had either gone out of business or were struggling to stay afloat. To increase sales, many small and medium-size breweries pursued a popular marketing strategy: releasing multiple cans with similar designs, in various colors.

This series by the Pittsburg Brewing Company stood out on shelves and encouraged people to collect the entire set.

Pittsburgh Brewing Co.
Pittsburgh, PA | 1970s

Pittsburgh Brewing Co.
Pittsburgh, PA | 1970s

Pittsburgh Brewing Co.
Pittsburgh, PA | 1970s

Reno Brewing Co., Inc.
Reno, NV | 1940s/1950s

Pittsburgh Brewing Co.
Pittsburgh, PA | 1970s/1980s

Menominee-Marinette Brewing Co.
Menominee, MI | 1950s

Boon Rawd Brewery
Bangkok, Thailand | 1990s

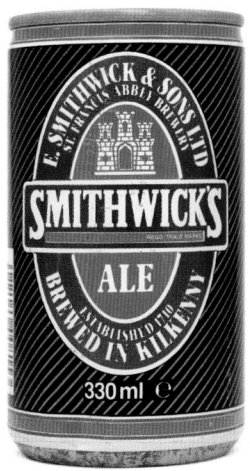

St. Francis Abbey Brewery
Kilkenny, Ireland | 1970s/1980s

South Pacific Breweries
Boroko, Papua New Guinea | 1970s

Galveston-Houston Breweries Inc.
Galveston, TX | 1940s/1950s

Southern Select

BEER

BREWED AND PACKAGED BY
GALVESTON-HOUSTON BREWERIES INC.
GALVESTON, TEXAS

The George J. Renner Brewing Co.
Akron, OH | 1940s/1950s

Speight's Brewery
Dunedin, New Zealand | 1980s

St. Pauli Girl

Originating in the seventeenth century in Bremen, Germany, St. Pauli Girl's name is derived from St. Paul's Monastery—the location on which the St. Pauli Brewery was built.

In the 1800s, the brewery commissioned a local artist to create a label for the brand, and the result featured a traditionally dressed waitress carrying steins of beer. To this day, the waitress remains a signature characteristic of the brand.

St. Pauli Brewery
Bremen, Germany | 1970s/1980s

BREWERY COMPANY · BELLEVILLE, ILLINOIS

Stag

BEER

Extra Dry Pilsener

COPYRIGHT 1947 · INTERNAL REVENUE TAX

CONTENTS 12 FLUID OUNCES

Griesedieck Western Brewing Co.
Belleville, IL | 1940s

Standard Brewing Co., Inc.
Rochester, NY | 1940s/1950s

Star Brewing Co.
Boston, MA | 1930s/1940s

Stegmaier Brewing Co.
Wilkes-Barre, PA | 1970s

August Schell Brewing Co.
New Ulm, MN | 1970s

Artois Brasseries
Leuven, Belgium | 1970s

STEGMAIER
Gold Medal
BEER

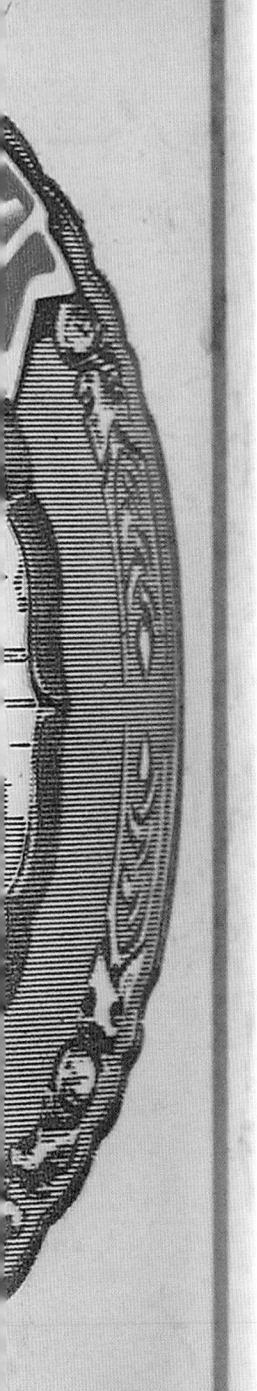

Sterling Kentucky Derby

Another multi-can series targeted to collectors, this collection by Sterling depicts the winning horses of the Kentucky Derby, one of the oldest and most prominent Thoroughbred horse races in the United States.

G. Heileman Brewing Co.
Evansville, IN | 1970s/1980s | front & back

G. Heileman Brewing Co.
Evansville, IN | 1970s/1980s

G. Heileman Brewing Co.
Evansville, IN | 1970s/1980s

G. Heileman Brewing Co.
Evansville, IN | 1970s/1980s

G. Heileman Brewing Co.
Evansville, IN | 1970s/1980s

Stern-Brauerei
Cologne, Germany | 1970s/1980s

Jones Brewing Co.
Smithton, PA | 1960s

Swan Brewery
Perth, Australia | 1970s

Rainier Brewing Co.
San Francisco, CA | 1940s

Carson Brewing Co.
Carson City, NV | 1930s/1940s

Taiwan Tobacco & Wine Monopoly Bureau
Wujih, Taiwan | 1980s

American Brewing Co.
Rochester, NY | 1940s/1950s

Tasmanian Breweries Pty. Ltd.
Tasmania, Australia | 1970s/1980s

Cerveceria Cuauhtémoc. S.A.
Monterrey, Mexico | 1970s

Pittsburgh Brewing Co.
Pittsburgh, PA | 1970s/1980s

NT CALEDONIAN BREWERIES LIMITED, SCOTLAND.

Tennent's Lager

While many breweries have adorned their cans with pictures of women, few are as well known for doing so as Tennent's. Known as the "Lager Lovelies," Tennent's featured various girls on its cans from 1962 through 1991.

Tennent's beer was very popular in its time, particularly within the military, and the cans remain popular among collectors today.

Tennent Caledonian Breweries Ltd.
Glasgow, Scotland | 1970s | front & back

Malayan Breweries Ltd.
Singapore, Malaysia | 1970s/1980s

Time Brewing Inc.
Dallas, TX | 1940s/1950s

Toohey's Ltd.
Sydney, Australia | 1970s/1980s

Rochester Brewing Co.
Rochester, NY | 1950s

Cerveceria Moctezuma S.A.
Orizaba, Mexico | 1960s

Tsingtao Brewery
Tsingtao, China | 1970s/1980s

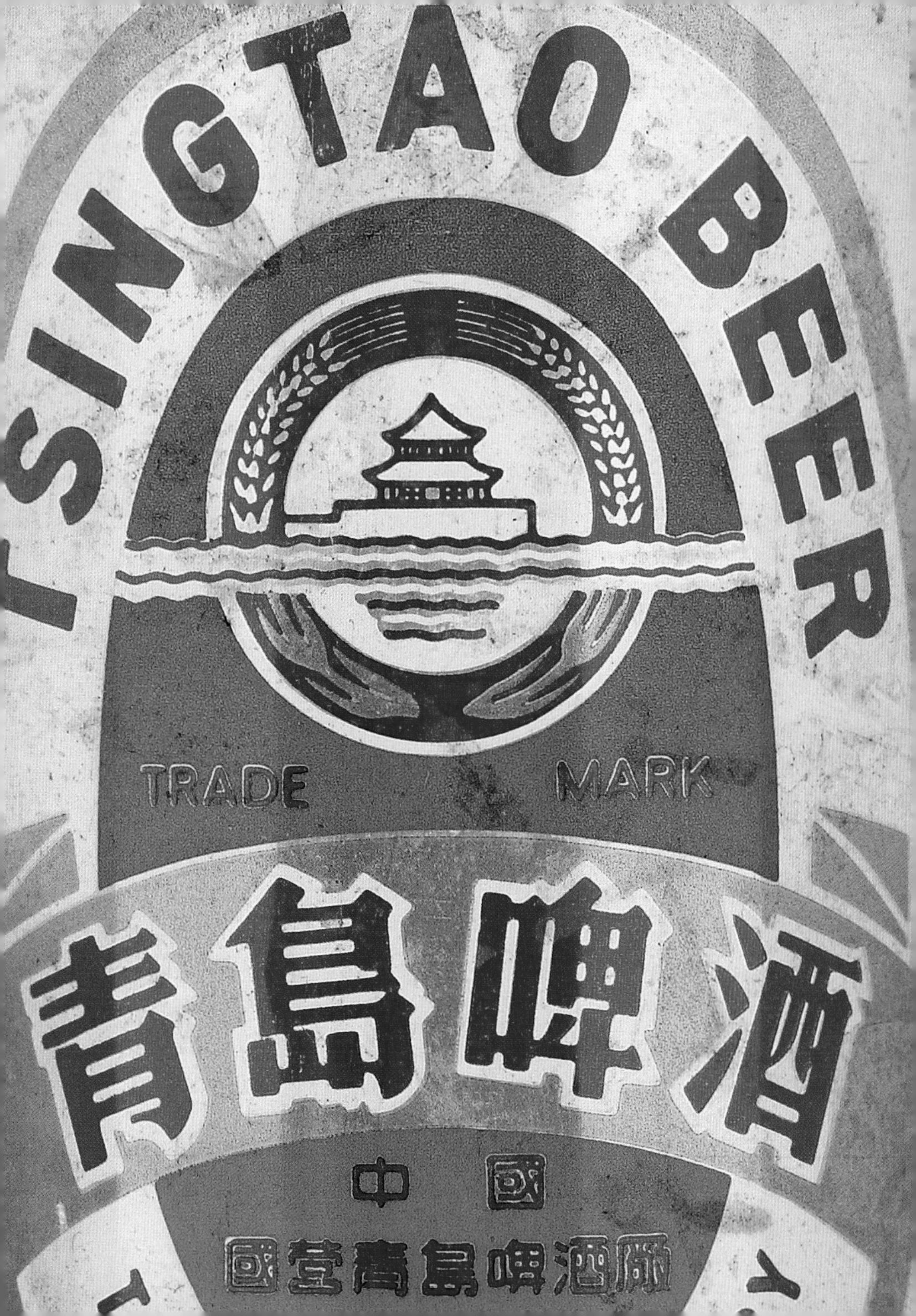
SINGTAO BEER
TRADE MARK
青島啤酒
中國
國營青島啤酒廠

Tudor

Produced by numerous breweries throughout the United States, Tudor was the store brand for The Great Atlantic & Pacific Tea Company, an American supermarket chain better known as A&P.

Metropolis Brewery Co.
Trenton, NJ | 1950s

West End Brewing Co.
Utica, NY | 1950s

El Dorado Brewing Co.
Stockton, CA | 1930s

C. Schmidt & Sons Inc.
Philadelphia, PA | 1950s

Van Merritt Brewing Co.
Chicago, IL | 1960s

Brewery Management Co.
New York, NY | 1940s/1950s

General Brewing Co.
Los Angeles, CA | 1970s/1980s

Van Merritt
Beer
Brewed from Imported
Based on An Old World
Formula, Originated in
1433 and Improved
Through the Centuries.
RED BEER
CONTENTS 12 FLUID OUNCES • FULLY PASTEURIZED

Walter Brewing Co.
Eau Claire, WI | 1950s

Joseph Huber Brewing Co.
Monroe, WI | 1970s

White Label Brewing Co.
Minneapolis, MN | 1960s

The George Wiedemann Brewing Co.
Newport, KY | 1940s/1950s

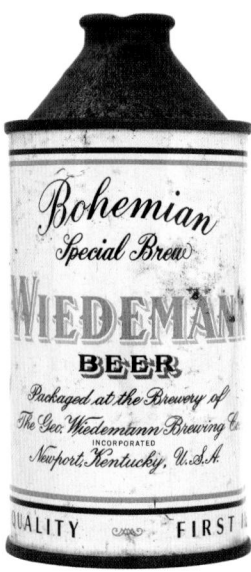

The George Wiedemann Brewing Co.
Newport, KY | 1950s

Pacific Brewing & Malting Co.
San Jose, CA | 1930s/1940s

The South West Breweries Ltd.
Windhoek, Namibia | 1970s

Joseph Huber Brewing Co.
Monroe, WI | 1960s

The Wooden Shoe Brewing Co.
Minster, OH | 1940s/1950s

Dalla Wührer
Brescia, Italy | 1970s/1980s

BREWED ONLY FR
Wunder
DAS TROCKEN-
Now brewed in Cincin
... by its Original
Braumeister
CONTENTS 12 FL

Wunderbräu

This premium German beer was designed to appeal to the largely German population of Cincinnati, Ohio, in the mid-1950s. Even the beer's slogan, "Das Trocken-Lager" (The Dry Lager) was printed in German on the can. Wunderbräu became so successful that its maker, Red Top Brewing, renamed itself after the product to Wunderbräu Brewing.

Wunderbräu Brewing Co.
Cincinnati, OH | 1940s

Würzburger Hofbräu A.G.
Würzburg, Germany | 1970s

D.G. Yuengling & Son Inc.
Pottsville, PA | 1970s/1980s

D.G. Yuengling & Son Inc.
Pottsville, PA | 1980s